Victorian and Edwardian

Windmills
and
Watermills

from old photographs

1   The Windmills of Montmartre photographed by Hippolyte
Bayard in 1842.

Victorian and Edwardian

Windmills and Watermills

from old photographs

Introduction and commentaries by

J. KENNETH MAJOR
MARTIN WATTS

FITZHOUSE BOOKS, LONDON

Phototypeset by Tradespools Ltd, Frome, Somerset
Printed by The Bath Press, Avon
for the publishers Fitzhouse Books, an imprint of
B.T. Batsford Limited, 4 Fitzhardinge Street
London W1H 0AH
First published 1977

ISBN 0 7134 0621 6

Contents

Acknowledgements

Especial thanks are due to the Society for the Protection of Ancient Buildings and to Alan Stoyel for help in tracking down photographs.

The photographs in this book appear by courtesy of the following: B. T. Batsford Ltd, 3, 8, 12, 20, 21, 22, 23, 27, 32, 33, 36, 37, 40, 41, 42, 43, 49, 57, 58, 63, 70, 89, 112, 113, 124, 128, 134, 135, 143; Birmingham Reference Library, 52; Bristol City Museum, 91, 92; Coulthard, A. J., 9, 24, 34, 39, 66, 136; Davies-Shiel, M., 67; East Suffolk Record Office, 10; Essex Record Office, 115, 117; Goonveen & Rostorwrack China Clay Co., 76, 77; Hereford County Library, 109; Institute of Geological Studies, 78, 79; Mugford, D., 71; Museum of English Rural Life, 104, 105, 106, 110, 111; Noad, J. & J. Millers, 100; Quick, Roy, 16; Sayer, Lady Sylvia, 69; Shove, Godfrey, 2; Somerset Archaeological and Natural History Society, 62, 94; Smithsonian Institution, 93; Smith, R. W., 114; S.P.A.B., 100, 101, 102, 123, 133; S.P.A.B.: Collection of Miss E. M. Gardner, 44, 46, 53, 56, 59, 74; S.P.A.B.: Collection of E. Mitford Abraham, 4, 5, 17, 18, 28, 29, 95, 96, 125, 126, 127; Stoyel, Alan, 64, 81; Tritton, Paul, 60; Vosper, D. C., 65, 72, 87; Watkiss, R., 103; West Suffolk Record Office, 84; Weybridge Museum, Elmbridge Borough Council, 119; Wiltshire Archaeological Society Library, Devizes, 120, 121; Wolverton Archaeological Society, 26; Woolf, Charles, 61. From the authors' own collections, 1, 6, 7, 11, 13, 14, 15, 19, 25, 30, 31, 35, 38, 45, 47, 48, 50, 51, 54, 55, 68, 73, 80, 82, 83, 85, 86, 88, 90, 98, 107, 108, 116, 118, 122, 129, 130, 131, 132.

Introduction

Windmills and watermills existed in large numbers throughout Britain up to the time of the 1914–18 war. The period covered by the photographs in this book is the time when the first signs of decline were becoming apparent and when the competition from the large steam-driven roller flour mills at the ports was causing mills to close down and to disappear. While there were several thousand of these mills at work at the beginning of this century, there is today only one windmill working for the support of its operators and a dozen or so watermills.

A windmill at work on a fine summer day has little to rival it for sheer beauty. It falls into the same class as a full-rigged ship at sea. Although many of the machines illustrated here are purely utilitarian, even in these there is a magic which catches the spectator's eye.

The machinery of windmills and watermills is fascinating; but it is hard to realise now just how much industrial work was done by the mills, especially the watermills. The section on mills in mining and industry shows a few of these industrial uses. Windmills were used to crush oil seeds and to power wood saws, but the watermill continued as the main source of power in the metal-mining areas long after the steam engine was introduced to the industrial scene. Areas remote from coal were forced to use water power, and so many sites, like that in the Coniston Fells (67), had waterwheels to power ore-dressing machines and to raise ore and men from the shafts. Waterwheels became extremely powerful and very large, as the Laxey Wheel (68) shows, to deal with the problems of the mining industry. The use of power in this way came to an end when electricity was distributed to the remoter parts of the country. The science of using water power to its greatest advantage developed in the latter half of the nineteenth century and is shown here.

The earliest known photographs of windmills were those taken by Hippolyte Bayard in 1842 of Montmartre. From that time windmills and watermills attracted the attention of photographers as they had the artists of the preceding century. The great photographers of landscape subjects in nineteenth-century Britain—Francis Frith, H. W. Taunt, and Frank Meadow Sutcliffe to name only three—all have a large number of mill photographs in their archives. The introduction of the camera as a means of newspaper reporting in the period covered by this book means too that there was much interest in photographs showing disasters, such as fires in mills. Local photographers also aimed to present views of their home towns or villages, and windmills and watermills were included in these collections.

Flashlighting in photography was only just coming into use, so photographs of the interiors of the mills and of millers at work are few. There are one or two interiors in this collection which are extremely good: that of Bossava Mill by one of the Gibsons of Penzance (103), and the picture by Miss Wight of the miller dressing his stones (109).

The end of the nineteenth century and the beginning of the twentieth century saw the development of the windmill into the form of the wind engine with annular sails. This form had a short life owing to the development of universal electricity for power and the introduction of the portable internal combustion engine. Examples of these are included as they are part of the development of wind power. Although they only lasted for a short time they are now coming back into use in view of the expense of other forms of power for pumping water. These machines also had an important part to play in the agricultural economy of the time.

Local windmills or watermills played a great part in the village economy, and this is reflected in many of the photographs. The mill was often a meeting place, and carts and farmers can be seen queueing up to deliver or to receive their grain or meal. Millers were largely proud (and wealthy) members of the community and the beauty of their mill was part of that pride to be reflected in a photograph.

Tribute should be paid to two assiduous collectors of mill photographs: Miss E. M. Gardner and Mitford Abraham whose material was left to the Wind and Watermill Section of the Society for the Protection of Ancient Buildings. There are other collections in archives in the country which remain to be investigated and recorded by mill enthusiasts.

The Illustrations

Windmills

2 Westfield Mill, Somerset. Latterly known as Heath House Mill, this typical Somerset tower mill was tail-winded and ceased to work in 1900, dating this photograph probably to the 1890s. The short cylindrical masonry tower, thatched gable cap, with the tail-box containing the winding gear, and broad sails set with canvas on longitudinal bars, was familiar in Somerset and perhaps the survivor of the early tower mill form. Westfield Mill was built in the late eighteenth century and last worked with a single pair of over-driven stones. One of the men by the mill door is William Tucker, the last windmiller at Heath House, who died in 1902.

3 Waples Mill Farm, near Dunmow, Essex. This picnic scene of 1896 shows Isaac Mead and his family at dinner in the harvest field in front of the derelict post mill.

4 Freckleton, Lancashire. This un-
dated photograph shows a fine small
post mill in the Lancashire tradition.
The roundhouse is extended upwards
to the underside of the body of the mill.
The steps were raised by the lever on
the upper part of the ladder.

5 Orford Mill, Suffolk. This open post
mill shows the construction of the base
of a post mill very clearly. The castle
built by Henry II at Orford can be seen
under the fantail of the mill.

6 West Ashling, Sussex. This is a combined windmill and watermill which had ceased to work by 1930. The watermill had three pairs of French stones on the stone floor. The windmill is a hollow post mill which drove a further three pairs of stones. The drive went down the centre of the main post. The ring below the buck is a reefing stage for the windmill sails. The fantail turned the mill into the eye of the wind by means of an elaborate gearing on a ring at the head of the post.

7 The post mill, granary and tide mill at Walton-on-the-Naze, Essex. The whole site was demolished in 1921. The tide pond was 30 acres in extent and drove one undershot wheel with straight radial floats.

8 Ashcombe Mill, Lewes, Sussex. The fine six-sweep post mill which stood on Kingston Downs, and was blown down in the March blizzard of 1916. The six patent sails, an unusual number for a post mill, were fixed to the windshaft by a three-way poll end, another rarity, if not unique, and the neck bearing of the mill was, at one time, of glass.

9 Hornchurch Mill, Essex. This post mill has now been demolished. In this photograph it is at rest with the shutters of the sails open in a St Andrew's cross. The mill had a tall buck on a two-storey roundhouse. The mill was turned to the wind by means of a tail pole. Note the sack slide down the tail ladder.

10 Post mill at Stoke, Suffolk, photographed about 1884. Although built originally in 1746, this presents the form of a nineteenth-century mill. It survived until 1887 when it was demolished.

11 Winchelsea, Sussex. A tall-bodied post mill with a roundhouse, with two common and two spring sails. This combination of sails gave a good working partnership of power with a certain amount of self-regulation, and the high, eight-bladed fantail on the back of the buck which drove to wheels on the foot of the ladder, kept the mill to the eye of the wind. In this early twentieth-century picture the mill is already showing signs of decay in the spring sail shutters.

12   Clayton Mills, Sussex, *c.* 1910. These two mills stand on the north facing slope of the South Downs. The tower mill, 'Jack', is now without its sails or gear. The post mill, 'Jill', was originally at work in 1821 in Dyke Road, Brighton, and was brought up to Clayton in 1848. The mill was brought up complete on drays drawn by oxen.

13 Windmills at Outwood, Surrey. The post mill has been preserved but the smock mill collapsed in the 1960s. The post mill was built in 1665 and it is reported that the Great Fire of London was visible from this mill.

14 The windmills at Great Hormead, Hertfordshire. The two mills are both at work at the time of this photograph. Now only the cellar under the smock mill and the post of the post mill remains.

15 Tower mill at Wycombe Heath, Buckinghamshire. This mill must have been close to failure when this photograph was taken at the beginning of this century as the structure collapsed in the 1920s. The deliberate posing is unusual in a postcard.

16 Vale Mill, Somerset. Built early in the nineteenth century, Vale Mill, on Worle Moor behind Weston-super-Mare, worked for about 100 years. The four common sails drove two pairs of millstones and the mill is seen here working under her last miller, Thomas Quick. She had stopped work by 1910, and was finally gutted by fire in 1962.

17 Brynteg Mill, Anglesey. The mills of Anglesey have all been reduced to shells and have lost their sails. This photograph of 1903 shows a tower mill with a steep batter, a cap turned into the wind by an endless chain and four common sails.

18 Great Crosby Mill, Liverpool. This mill was already out of use when this photograph was taken in 1900. The steam engine in the mill yard may have been the reason for the abandonment of wind power.

19 St Davids, Pembrokeshire. The last windmill to work in south-west Wales was built in 1806 above St Davids and worked until 1904. The cylindrical, squat tower, with its broad sails, owes nothing to the windmilling improvements which can be seen in the East Anglian and home counties windmills dating from that time. The bowsprit, extending forward of the windshaft, is more reminiscent of the mills of the Iberian peninsula. In 1907 the windmill was converted into an hotel, and this illustration dates from about that time.

20 The tower mill at Buxhall, Suffolk, shown at work in this photograph of about 1890. This mill was built by William Bear of Sudbury. On the back of the photograph is the following note by N. G. Etheridge an employee of the mill: 'The mill is where I used to work, the chapel is where I used to preach, the old house that is falling down is where I used to work before God called me to His work 29 years ago.'

21 The tower mill in Ditchling Road, Brighton, about 1893. This mill, which was built in 1838, was demolished in 1913. The millers were C. Cutress & Son whose shop is to be seen in the adjacent house. One interesting point about this mill is the circular extension for storage which has been built around the base of the tower.

22 High Mill, Great Yarmouth, Norfolk. This mill was built in 1812 and demolished in 1905. It was the largest windmill ever built in England. The tower was 102ft to the cap, the diameter of the mill at ground level was 40ft and the walls were 3ft thick. The lantern on top of the cap was 122ft from the ground.

23 Patrington Windmill, Yorkshire, about 1900. This large mill shows many interesting points of windmill construction. The fantail is carried high above the cap. The mill is at rest with the canvas-covered shutters open. The way in which the sails curve from root to tip is shown well in this picture.

24 Southsea, Portsmouth, Hampshire. The Dock Mill at Southsea was built by the Dockyard Co-operative Mill Society, formed by shipwrights to help the poor in the early nineteenth century when bread prices were very high, in about 1817 and was worked by them until 1834. After being disused for some years the mill was purchased by Maurice Welch, a miller, in 1869. In 1900 the millstones were replaced by a steel roller mill and wind power was last used in 1905. This picture must date from about that time, as the sail shutters have all been removed. The big black mill in its urban setting was finally demolished in 1923.

25 Riddings Mills, Derbyshire. These mills were both built in 1877 and were known as 'Sarah' and 'James' after members of the owner's family. Both are now shells. The six sails were supported on heavy iron windshafts with 'crosses' to which the sails were bolted.

26 Great Horwood Windmill, Buckinghamshire, in 1910. Although the sails are already rather derelict, the fan tail is rotating to turn the mill into the wind. A steam engine must have driven the mill, for there is a pulley wheel at the head of the ladder at first-floor level. This will have driven the mill through the crown wheel on the upright shaft.

27 Union Mill, Whitby, Yorkshire. This communal mill dates from about 1800 when the Napoleonic wars drove the price of flour and bread to almost astronomical heights. Similar communal mills which operated for the flour consumers' benefit until the end of the nineteenth century became an integral part of the Co-operative movement.

28 Bromborough Mills on the Wirral Peninsula. This photograph of 1870 shows the fine white-painted tower mill built above the steam and water-mills.

29 Alderton Mill, Suffolk. This photograph shows the fine smock mill which has four shuttered sails. There is a portable steam engine in the mill yard together with a threshing machine and an elevator. Were they there to thresh the corn from the mill or was the miller a harvesting contractor?

30 The tall windmill at Berney Arms on the River Yare in Norfolk, c. 1890. The semicircular casing to the right of the mill contains the 18ft diameter scoop wheel for lifting water to drain low-lying land.

31 Flour mill at Wainfleet near Skegness, Lincolnshire, in 1915. This is one of the typical 5-sailed mills of Lincolnshire.

32 The windmill at Lewes prison, *c*. 1875. This is a case of a windmill which has been removed from one site to another. This smock mill was first erected at Pipes Passage, St Michaels, Lewes, and was moved a few years later to the prison site. The sweeps and fan were removed in 1912 and the mill was demolished in 1922.

33 Willesborough Mill, Kent. This smock mill was built by John Hill of Ashford in 1869 for Cornes and Sons. When driven by the wind up to four pairs of stones could be used. Further stones were driven by the steam engine and later by electric motors.

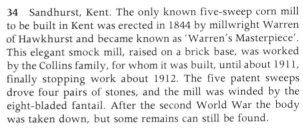

34 Sandhurst, Kent. The only known five-sweep corn mill to be built in Kent was erected in 1844 by millwright Warren of Hawkhurst and became known as 'Warren's Masterpiece'. This elegant smock mill, raised on a brick base, was worked by the Collins family, for whom it was built, until about 1911, finally stopping work about 1912. The five patent sweeps drove four pairs of stones, and the mill was winded by the eight-bladed fantail. After the second World War the body was taken down, but some remains can still be found.

35 Leach's Mill, Lynn Road, Wisbech, Cambridgeshire. A photograph taken before 1895. This very tall 8-sailed windmill is shown at work. The sails were mounted on the cast-iron 'cross' at the end of the windshaft and the shutters were controlled by the striking rod coming out of the centre of the shaft. This mill was supplemented by the adjacent steam mill.

36 Ibstone Mill, Buckinghamshire. This small smock mill stands on a prominent ridge in the Chiltern Hills, and is now undergoing preservation. It has two pairs of stones.

37 Old Trumpington Windmill, Cambridge, about 1880. This brick tower mill was built in 1812 and was demolished about 1890 having ceased to work in 1887. In this photograph it is at rest with the shutters open.

38 Biscot Windmill, near Luton, Bedfordshire. This smock mill has now gone. It was a very broad-based mill, but at the time of this photograph it had been reduced to only two sails.

41 Ballards Mill, Patcham, north of Brighton, *c.* 1870. This smock mill was built in 1791 and demolished in 1902. The method of controlling the sail shutters from the spider at the end of the windshaft can be seen.

39 Ramsgate, Kent. A fine Kentish smock mill built early in the nineteenth century by a Mr Knott, and considered a well-found mill, four patent sweeps driving four pairs of stones. By the early 1920s only the base survived as a motor garage, and a powerful illustration of the sad remains was made by Frank Brangwyn. Even the base has now gone.

40 West Blatchington Mill, Hove, about 1880. This smock mill was built on a square base from which three barns radiated. It is now preserved by Hove Corporation.

Watermills

42 Felsted, Essex, the mill in about 1890. This late nineteenth-century mill has an enclosed waterwheel and a further weather-boarded skirt hangs down over the tail race. The miller stands with arms akimbo by the escape sluice.

43 The tide mill at Bishopstone between Newhaven and Seaford, Sussex in August 1883. The windmill surmounting this mill is believed to have been used solely to power the sack hoist. The tide mill was operated by a famous miller named William Catt and it was he who added the windmill. The mill pond was formed from pounds fed by cuts from the river Ouse. These in turn drove three 15ft diameter undershot wheels and sixteen pairs of millstones.

44 Holy Street Mill, Chagford, Devon, a picture dated 1856. This wooden waterwheel was replaced by an iron wheel soon after this picture was taken. There are several important watercolours of this mill including those painted in the 1860s by Miles Birkett Foster.

45 Upper Middle Mill, Uplyme, Dorset. This postcard shows a very small West Country mill at the turn of the century. The wooden overshot wheel drove only two pairs of stones.

46 Gatewick Mill, Sussex, which was demolished in 1878. The cart with a canvas tilt in the lean-to shed is a typical miller's cart.

47 Hele Mill, near Ilfracombe, Devon, about 1900. This shows the mill at work with water discharging down a sloping launder on to the overshot waterwheel. The miller is at the door with a sack. This obviously posed group is in its best bonnets.

48 Manor Mill, Woolacombe, Devon. A fine overshot wheel for farm purposes, from which the drive was taken by an external ring gear, just visible on the side of the wheel nearest the building. The lady sitting on the wheel-pit wall in this nicely-posed picture holds the outer end of the shaft which carried the pinion driven by the ring gear. The rounded building to the left perhaps contained a horse gear.

49 Lords Mill, Chesham, Buckinghamshire, taken *c.* 1890 by W. Butts, a local photographer. This mill has had a succession of waterwheels. At the time of this photograph this external wheel was added and one of the two internal waterwheels was discontinued. Now all that remains of this fine mill is one internal waterwheel and the building.

50 The old watermill, Totnes, Devon. This postcard, dating from before 1908, shows the breast-shot waterwheel rotating. The timber structure to the left of the wheel is a dust chute taking the chaff away and discharging it into the stream.

51 Mapledurham Mill, Oxfordshire. This mill is one of the Thames mills and only stopped working in the late 1930s. In this picture, published in 1895, the mill is shown with two low-breast-shot wheels. Frank Frith & Co. photographers.

52 (overleaf) The Old Town Mill, Tredington, Warwickshire, in 1896. This photograph shows the tail race being used as a sheepwash by local farmers. The large waterwheel is an all iron wheel which is breast shot. The pulley wheel on the right implies some form of outside drive from a steam engine or a gas engine.

53 Havant Mill, Hampshire, a good building with mansard roof dating from the early part of the nineteenth century. This form of breast-shot wheel and its position in relation to the building is typical of this part of Hampshire.

54 East Mills and Bridge, Colchester, Essex, about 1905. This postcard shows a Thames barge at the mill which was both a steam mill and a watermill. Large flour mills such as this show their many stages of growth in the development of the building.

55 Strand Mill, Dawlish, Devon. This 30ft diameter waterwheel was built by A. Bodley of Exeter. The water enters high up on the wheel at the '11 o'clock' position. The launder is carried on the brick and stone pillars from the mill pond on the bank behind. This was a corn mill.

56 Brantham Mill on the river Stour on the Essex-Suffolk border. This picture shows a fine trading mill which, to judge from the number of carts and the Thames barge, was extremely busy. A steam engine was later installed to assist the waterwheel.

57 Heybridge Mill, Essex, a photograph by Frank Frith & Co. This large watermill was at work when this photograph was taken in 1900. The double doors in the centre of the picture lead on to the wheel. The various lines in the weather-boarding and in the brick base show how the mill has been extended over the years.

58 Newark Mill, near Ripley, Surrey. A photograph by Frank Frith & Co., dated 1903. This great mill was burnt down in 1963 although its preservation had seemed assured. The mill is on the Wey Navigation, hence the projecting door on the left at first-floor level, and was powered by three waterwheels.

59 Langford Mill near Maldon, Essex, which was burnt down in 1879. This picture is interesting because it shows the number of workers in the mill and a donkey cart below the sack door on the first floor.

60 Hare Street Mill, Great Waltham, Essex. This photograph of about 1900 shows a good weather-boarded water-mill. The two lucams and the walkway around the roof form a good example of the accidental design of satisfactory industrial buildings in the functional tradition.

61 Copperhouse Tide Mill, Hayle, Cornwall. The principle of the tide mill was to use water impounded at high tide in a large pond and released as the tide ebbed. At Hayle the tide mill was driven by a 20ft diameter waterwheel, and also by a steam engine when the water was insufficient, for as the level in the pond fell, the working head obviously became less and less. Also hours of working, although predictable, were irregular and it is said to be for this reason that the Copperhouse Mill eventually closed. The steam engine stack and a small building can still be seen at the site of this former corn mill.

62 Bathampton Flour Mill, which stood at the south end of a weir across the river Avon, east of Bath. Like several of the weirs along the Avon, there was a mill at each end and the ferry slip, which was put out of use by the building of a new bridge in 1871, can be seen to the right of the mill. The mill itself was rebuilt after a fire of 1861, when the miller was Thomas Spackman. The house attached to the rear of the mill still survives, but of the mill only the wheel arch remains at the end of the weir.

63 Tide mill at St Osyth, Essex. This mill was still at work when this potograph was taken in 1910. It has now been demolished. The loading and unloading arrangements at the mill are of interest. Barges such as this hay barge could be loaded at the rear door of the mill in the extension. Carts were unloaded at that end and the sacks were drawn up into the projecting lucam in the gable.

Mills in Mining and Industry

64 Glynn Valley China Clay Pit, Cardinham, Cornwall. Water power was widely developed and used in Cornish industry and this photograph of *c.* 1870 shows a fascinating combination of functions. The 35ft wheel was at the same time both overshot and breast shot, water being brought to it by wooden launders. The streamlined effect at the end of the upper launder was to prevent the water being blown off the top of the wheel in strong winds. On the near side, the man, one H. Bilkey, has his hand on a rod connected to a crank on the wheelshaft. This rod was connected to a balance box, for the rods leaving the picture to the bottom right drove a pump remote from the wheel, for lifting china clay slurry. On the far side of the wheelshaft is a winding drum, which would have carried a cable for haulage in the clay pit.

5   A Pelton wheel driving small stamps, set up under the
cliffs north of Cape Cornwall and photographed about 1900.
The Pelton wheel, set up in the left foreground, is a form of
impulse turbine developed in America in the second half of
the nineteenth century. The wheel revolved at a high speed,
turned by a jet of water supplied at great pressure, usually
from a high head. The water here was supplied by the pipe
coming down the cliff to the left hand side. The drive was
taken from the wheel to the compact set of stamps by belt.

66 Punnetts Town Mills, Sussex. Wind-powered sawmills were generally uncommon in England and one of the last mills to drive saws was brought to Punnetts Town in the 1860s. The octagonal smock mill raised on a timber base drove three circular saws, two on the ground floor and one on the first. The saw-mill was demolished in 1933.

67 The dressing works of the Coniston Copper Mines in the Lake District. This is one of the earliest photographs of mining in the Lake District and it was taken about 1860. The rotating waterwheel in the foreground is driving a series of vibrating tables to separate out the ore. The row of open sheds in the background were also supplied with power from waterwheels built between each pair. The wall behind was the storage pound for the water.

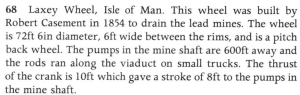
GREAT WHEEL

68 Laxey Wheel, Isle of Man. This wheel was built by Robert Casement in 1854 to drain the lead mines. The wheel is 72ft 6in diameter, 6ft wide between the rims, and is a pitch back wheel. The pumps in the mine shaft are 600ft away and the rods ran along the viaduct on small trucks. The thrust of the crank is 10ft which gave a stroke of 8ft to the pumps in the mine shaft.

69 Whiteworks, Dartmoor. One of Robert Burnard's fine photographs of his native Devon, showing the breast-shot wheel and tin stamps at Whiteworks near Princetown. The works were in action from about 1820, but this photograph, taken on 1 June 1889, shows the stamps barrel dropped out of position and the stamping had obviously ceased.

70 The great waterwheel at Benthall Bank, Shropshire. This wheel was attached to the ironworks in Benthall Bank. A previous wheel for the blowing engines on the furnaces was installed in the 1780s.

THE OLD WHEEL, BENTHALL

71 Tregrehan Mills, St Austell. The high breast wheel on the right-hand side of this photograph taken *c.* 1900 originally drove a corn mill, but was adapted to drive three sets of Cornish stamps, the water also being used to wash the tin ore which came locally from a place known as Downpark or Boscoppa. The lifters on the stamps barrel can be clearly seen, although only the centre set of four stamps are in place. The owner, Mr Tregilgas, stands on the right.

72 Levant Stamps, Cornwall, photographed about 1896. One of the most famous Cornish mines, situated north of St Just in the far west of the county, showing the Cornish stamps floor with steam and water power at work together, indicating the great demand for motive power in nineteenth-century industry.

73 Poltesco, Cornwall. The factory which made objects of serpentine. This stone was cut with water-driven saws and then polished into slabs or made into ornaments such as stone eggs or lighthouses.

74 The gold-dressing mill, Tyn Y Gros, by the Mawdach estuary in Wales. This early postcard shows the wheel used for crushing gold ore in the gold-mining field near Dolgellau.

75 Lee Moor, Devon. An overshot waterwheel at work, driving a sand grading and washing plant at the Lee Moor Brick and Tile Works. The photograph, taken by T. C. Hall in July 1908, was one of several taken by him of water-powered West Country industrial plants.

76 Tregargus Mills, Cornwall. This fine photograph by Burrows of Camborne, taken in 1896, shows the newly-erected waterwheel and the foundations for No. 2 mill at Tregargus. The wheel, 30ft diameter by 5ft wide, was built by F. Bartle & Sons of Carn Brea, and is dated 1896 with the maker's name on the shroud. The bowler-hatted man with his foot on the wheelshaft is Charlie Bartle. The wheel shows the high degree of technical refinement achieved by the end of the nineteenth century and appears to have been coloured, the panels on the shrouds being lighter than the arms and junction plates. Of note also is the trolley containing china stone above and to the right of the wheel.

77 Tregargus Mills, No.1 Mill. This wheel was also built by the Carn Brea Foundry in 1896, and was 22ft diameter by 7ft wide. Seen beside it is Charlie Bartle, with suit and stick, and two workmen. This wheel, in the lowest of the five Tregargus mills, drove four pans for crushing china stone.

78 Tregargus Mills. The exterior of No.2 mill, looking north east, with mills 3 and 4 behind. This remarkable series of five mills were built to use the natural fall of water on the site, the tailrace of the top mill feeding the wheel of the mill below and so on, down to No. 1 mill, the lowest. The wheel of this mill is that shown after erection but before the building was completed above. In 1968 the wheel was saved from destruction by scrap merchants as it was about to be blown up.

79 The interior of one of the Tregargus china stone mills showing the grinding pans, taken by T. C. Hall in 1905. These pans were under-driven by bevel gearing and the four projecting arms carried adjustable wooden hanging arms which simply pushed the lumps of china stone around in water. The lumps were gradually worn down, and the ground material removed from the pan as slurry. The five mills at Tregargus drove a total of nineteen pans between them.

80 Delabole Slate Quarry, Cornwall. This deep pit for the quarrying of slate was powered at first by the pumping waterwheels shown in the foreground of this postcard. The water from the tail of the first wheel drove the second wheel.

81 Although marked 'Tolvadon Tin Streams, Tuckingmill', this photograph shows in fact the Tolgarrick Tin Streaming Works, near Camborne in Cornwall, looking downstream. In the left foreground the overshot wheel is driving a dipper wheel, raising the tin ore in solution, and a small breast wheel in the middle of the picture is probably working equipment to sort or grade the ore.

82 Willsbridge, South Gloucestershire. This mill was built by John Pearsall in about 1716 for rolling iron and making steel, but the premises were sold and converted for corn milling after a patent taken out for iron roofing had failed. This was one of the many enterprising industrial ventures which grew up along the Avon valley in the eighteenth century, and the building and mill pond still survive.

83 A photograph of Bourton Foundry in Dorset dating from the period prior to 1914. The great wheel on the right of the photograph was 60ft diameter by 2ft wide. This pitchback wheel suffered from a shortage of water. It drove machinery in the foundry of the engineers and millwrights Maggs and Hindley.

Wind Engines

84 Bury St Edmunds, Suffolk. This wind pump was built by John Wallis Titt of Warminster for the Bury St Edmunds Corporation in 1898. This was the largest version of the firm's 'Simplex' engine. The wind wheel was 40ft diameter, the 50 sails were each 12ft long, 2ft 3in wide at the rim and 1ft wide at the inner end. These were adjustable according to the windspeed. The whole was turned into the wind by the two fantails visible in the centre. The tower was 35ft high and cost in all £550.

85 A wind engine and pump house built by John Wallis Titt at Southport about 1905. This engine lifted water into the large tank for the purposes of street washing and flushing sewers.

86 A farm windmill at Angmering, Sussex. This mill was built in 1853 by a member of the Warren family and it survived until about 1930. The eight sails were adjusted by the mechanism in the centre of the sails. In this photograph the fantail is missing.

87 Bollowall Farm, St Just, Cornwall. Although in its traditional form for corn-milling the windmill was not widespread in Cornwall, water-power being generally well-developed, wind-power was proposed and used for pumping, both in the mining industry and on a domestic scale. This small, probably home-made, wind wheel, set up between two farm buildings to catch the wind, was used to pump water. The photograph was taken about 1910; by 1920 only the cross timbers which supported it survived.

88 The windmill at Charlbury, Oxfordshire, a postcard printed before 1905. This shows a wind engine for pumping purposes mounted on a wooden tower. The make is not known but it appears to be one of the adjustable-bladed type.

89 Tenterden, Kent, a photograph of about 1890. This is an important early example of a wind engine in use for pumping water. It is not possible to determine the make nor the purpose of the intermediate gears in the centre of the wooden tower.

90 The windmill at Haverhill, Suffolk. This mill was erected for Richard Raffles about 1855. The tower was 60ft tall and the annular sails 50ft diameter. Although not the only corn mill of this type it was the most famous as it was not demolished until the 1939–45 war. This was done to prevent damage to aircraft.

91 The three-masted barque *Amity* in Cumberland Basin, Bristol. Built in 1877 she carried timber from the Baltic and between her main and mizzen masts is a 4-sailed wind-pump, a typical feature of many of these timber barques, which earned them the nick-name 'Onkers' from the distinctive sound made by the clank of the pumps. When not in use the sails were furled back along the arms to the centre.

92 *Topdal* being towed into Cumberland Basin, Bristol, a three-masted Scandinavian timber barque with 6-sailed wind-pump between her main and mizzen masts. Often old or leaky by the time they were 'relegated' to the timber trade, these three-masted vessels needed the constant use of their wind-powered pumps.

93 A view of a portion of the showground of the Worlds Columbian Exposition in Chicago 1893. Every possible form of wind engine is on display from the various types of fixed bladed mill to the adjustable bladed mill such as the Halladay which appears to be the large one in the centre.

Mill Machinery

94 Blackmore Manor Mill. This quiet interior of a small West Country corn mill shows clearly the arrangements by which the corn was taken into the eye of the stones. The sacking chute leading into the hopper, supported by the horse with its finely turned legs. The turned bar across the horse provided fine adjustment to the crook string, which regulated the angle and direction of the shoe. Of note also is the inverted wooden-cogged crown wheel, for an auxiliary drive to be taken off from the upright shaft, top left, and the rough, un-barked prop supporting the floor beam over the stones.

95 Keston Mill, Kent. An interior view of 1910, showing the brake wheel which is all wood, the brake and the one-piece cast-iron wallower at the head of the upright shaft. The millstones can be seen but they are without tuns, hoppers or shoes which shows that the mill was already out of use.

96 Keston Mill, Kent. A view of the floor below that shown on the previous picture. Here meal is received from the stones above and sorted into three grades by means of sieves.

97 Dunster, Somerset. The Lower Mill at Dunster, also known as the Castle Mill, still survives with two external overshot waterwheels, but in a disused state. This clear interior shows the mill at work early this century, the upright shaft rotating between the two pairs of millstones, and the sacks in the foreground were moved during the exposure.

98 Dunster, Somerset. This is a post-card by Frank Frith and Co. of about 1905. It shows the two outside water-wheels of this attractive mill both at work. Each wheel was coupled to two pairs of stones.

99 Although entitled Lamorna Mill, Penzance, this is in fact Bossava Mill. In this postcard one can see the small over-shot waterwheel and the peculiar cyclo-pean granite wall at this end of the mill.

LAMORNA MILL PENZANCE

100 Littleton Wood Mill, Wiltshire. An interesting photograph taken in 1887 of a waterwheel prior to its installation in a West Wiltshire corn mill. The wheel, 18ft diameter by 9ft wide, was built by G. Dunford & Sons, the Great Cheverell millwrights with castings from the Bratton Iron Works. The wheel cost £240 to install and in 1927 it was broken up and replace by a turbine, which was still in regular use in 1976.

101 A laboratory test of a small overshot waterwheel. This wheel is 6ft diameter and 1ft wide and is working against the water pressure of a pump through a crank and piston.

102 A wheel awaiting tests in the same laboratory. This larger overshot wheel with a gear ring some distance in from its rim has not yet been connected to the test equipment. It is not known where this testing is taking place.

103 Bossava Mill, Lamorna, Cornwall. This interior shot taken with a flash exposure is by one of the Gibsons of Penzance and is believed to date from 1880. This excellent photograph shows very clearly the details of a simple mill. Here the waterwheel is connected by means of the pit wheel—the large wheel behind the sack—to the single pair of millstones through a lantern gear. The meal discharges straight into a shaking sieve which separates it into its various grades. Behind the ladder there is the winch for lifting the sack to the upper level. There is no storage in this mill, the sacks were emptied one at a time into the hopper above the millstones.

Millers
and Millwrights

104 Synod Mill, Cardiganshire. Mr J. G. Jones is opening the outlet from the bin on the floor above and the grain will pour into the hopper and so between the millstones. The tun, or casing to the stones, the horse and the hopper are visible in this photograph.

105 Synod Mill, Cardiganshire. Mr J. G. Jones watches the crushed oats being discharged from the millstone chute into the bin to observe the quality of the meal.

106 Bagging grain. A pair of mill hands are measuring grain by the bushel and bagging it up. The bushel measure is being filled in front of a winnowing fan which was driven by hand.

107 Millstone Factory. This is part of the trade card of the Société Générale Meulière at La Ferte-sous-Jouarre (S-et-M) and it was used in 1905 to introduce their traveller Marcel Jaeck. This shows the mass production of French stones for the milling of fine flour. These stones were exported all over the world and were made from several pieces of hard natural stone made up of silica, carbonate of lime and other elements.

108 The millwright dressing stones. The millwright is dressing a French stone. The lands and furrows can be clearly seen. The millwright supports his left arm on a bag of chaff and is tapping the sharp steel mill bill on to the stone to produce the furrows in the harp which is the name for the triangular pattern of the furrows.

109 This remarkable photograph of a miller dressing his stones was taken by Miss Wight of Hereford. The runner stone of a pair of French stones has been turned to face upwards supported on bags of chaff. The miller has the thrift in his hands and he is supported on a cushion of chaff. The bill is wedged into the thrift. The bill is at work chipping furrows into the stone so that it can grind efficiently.

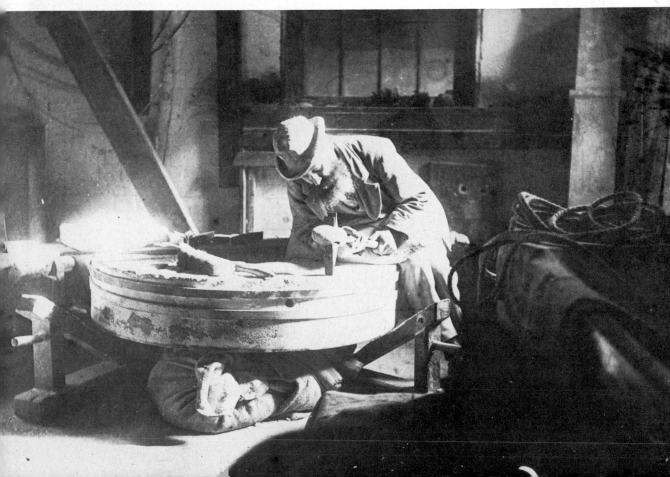

110 Moving the sacks about in the granary storey of a mill. This is a mill without bins in which the grain must remain in the sacks until it is ground. At this stage it would be poured down chutes into the hoppers of the millstones.

111 Turning grain in a kiln. This rather later photograph shows the miller at Hincaster in Westmorland turning oats which have been dried in the kiln over an anthracite fire. Kilns of this type only occur in the north of England and in Wales.

112 Joseph Bridge of Victoria Street, Bury St Edmunds, made this large waterwheel about 1899. The man on the left nearest the camera is holding a wooden gear tooth in his hand.

113 The millwright's yard. This is the yard of Medhurst's in Western Road, Lewes, Sussex, and this photograph dates from the 1860s. It is thought that the wheel in the yard was for the mill at Fletching which was demolished in 1950.

114 The miller's carts at Ongar Mills, Essex. The picture shows the office block on the main street from which seeds and meal were distributed in the four-wheeled carts. The coach-built bodies are unusual: miller's carts usually had canvas tilts. Perhaps this is a later version as the photograph dates from 1908.

115 Rochford, Essex, the miller's cart, a photograph of about 1905. Here the sacks are being slid from the first floor of the mill on to a cart. The carter is wearing a sack over his head for protection.

116 The Old Mill, Isleworth, Middlesex, about 1907. This postcard shows a Thames barge being unloaded at the mill below the three lucams. It is interesting to note that there is a sailcloth canopy attached to the lucams which would enable unloading to take place in bad weather. This was a large mill and here it is obviously receiving grain from the Port of London.

117 Bulford Mill, Cressing, Essex. A photograph of about 1900, showing the miller's carts setting out from this large watermill to deliver flour.

118 Samlesbury Mill, Lancashire. A group showing the workers in the mill in about 1910. The boss is obviously proud of his gun!

119 Coxes Lock Mill, Weybridge, Surrey, a photograph of *c.* 1889. This mill is still at work although it has changed out of all recognition. The mill staff are standing on the walkways above the barge *Perseverance*. This barge was built at Honey Street boat yard on the Kennet and Avon Canal.

Disasters
and Demolitions

120 Tilshead post mill, Wiltshire, photographed shortly before being pulled down in about 1905. This mill was one of the last surviving Wiltshire post mills and the thatched buck roof, vertically boarded roundhouse and wooden poll end are worthy of note. The mill was in the ownership of Jane Hussey in 1813 and was last worked by Thomas Long, the village baker and shopkeeper.

121 Tilshead Mill from the tail, showing the ladder and broken off tail-pole in an advanced state of decay. When the mill was pulled down, some of the timbers were used to rebuild the chancel roof of the village church. Several other windmills were situated on similar impressive sites overlooking Salisbury Plain.

122 The windmill at Freshwater, Isle of Wight, in 1869. This mill fell out of use when a steam-driven mill was erected beside it.

123 Ringmer, Sussex. This shows the wooden windshaft, the iron poll end which carries the sail and the neck bearing, the wooden brake wheel which drove the forward pair of millstones and the iron tail wheel which drove the back pair of stones.

124   The post mill at Wavertree, Liverpool, in 1892.

125 The post mill at Wavertree, Liverpool, in 1900.

126 The post mill at Wavertree, Liverpool, in 1902.

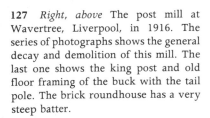

127 *Right*, *above* The post mill at Wavertree, Liverpool, in 1916. The series of photographs shows the general decay and demolition of this mill. The last one shows the king post and old floor framing of the buck with the tail pole. The brick roundhouse has a very steep batter.

128 The mill at Godmanchester, Huntingdonshire. This mill on the Great Ouse was derelict when this photograph was taken about 1900. There was an all-timber breast-shot waterwheel in the shaped structure on the left of the mill. The timber framing of the mill had decayed so that the whole building was twisting and settling into the river.

129 Iffley Mill, Oxford. The view prior to the fire of 1908.

130 Iffley Mill, Oxford. This mill was completely destroyed on 20 May 1908. Before its destruction it was frequently painted and photographed as one of the beauty spots on the Thames. A plaque now marks its site and dates the fire.

131 Iffley Mill, Oxfordshire. General view of the mill after being burnt out.

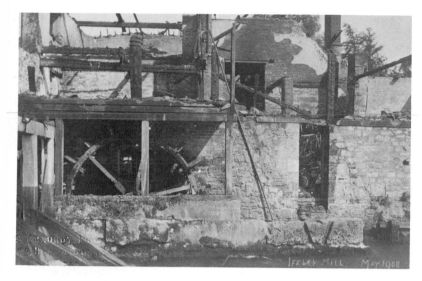

132 Iffley Mill, Oxfordshire. Detail of the wheel and by-pass hatches after the mill had been burnt out. The photograph, by G. Chaundy of Oxford, is dated May 1908.

133 Ringmer, Sussex. This shows the sails and the post and quarter bars taken during demolition. The post carried the buck of this post mill and the method of supporting the post can be clearly seen.

134 Layham Watermill, Suffolk, on fire about 1895. This fine mill which had a weather-boarded structure above the brick of the ground floor was still smouldering when this photograph was taken. The large iron wheel remains almost complete, but the gearing on the upright shaft has broken with the fire It is almost amusing to observe the equanimity with which the family posed for this photograph with their backs to the fire.

135 Hodsons Black Mill, West Hill Road, Brighton. This photograph was taken by Edward Fox on 25 June 1866 during the demolition of this 12-sided smock mill. The mill was built in 1804.

136 Warton, Lancashire. This derelict post mill, known locally as a 'peg' mill, was said to have been brought across the Ribble from Rufford in 1771, and, as photographed here, was the last remaining post mill in Lancashire. The mill contained only one pair of stones, reputedly 5ft 10in in diameter, and the lantern wallower, which can be seen above the decaying boarding, is of note. The tapering brick round-house enclosing the sub-structure also gave some support to the mill body.

137 Madingley, Cambridgeshire. A photograph of 1909 showing a post mill extremely decayed. This is not the preserved mill which stands near the American war cemetery. This photograph shows the relationship between the post, the quarter bars and the body or buck which rotates about the head of the post.

Glossary

ANNULAR SAIL   A single row of radial shutters in a circle.

BATTER   Slope or curve of the wall of a tower mill.

BEVEL GEARS   Gears profiled to drive at right angles, or nearly so, to each other.

BILL   Hard steel double-ended chisel used to dress millstones.

BRAKE WHEEL   The primary gear wheel in windmills, mounted on the windshaft, on which the brake acts.

BREAST-SHOT WHEEL   A waterwheel where the water enters at about the level of the wheelshaft.

BUCK   The body of a post mill, containing the machinery.

CAP   The moveable top of a tower/smock mill.

COMMON SAILS   The traditional cloth-set sails, where cloth is set on a lattice frame.

CROOK STRING   A cord used to adjust the angle and direction of the shoe.

CROSS   Four-, five-, six- or eight-armed casting mounted on the end of the windshaft to which the sail backs are bolted.

CROWN WHEEL   Commonly, the bevelled face gear on the upright shaft from which secondary drives are taken.

DIPPER WHEEL   A driven wheel with shaped buckets to lift tin ore in solution.

FANTAIL   A wind wheel set at right angles to the sails, which turns the mill into the wind automatically.

FRENCH STONES   Millstones built up with pieces of stone imported from the Paris basin, also known as French Burrs.

FURROWS   Grooves cut into the grinding faces of millstones.

GALLERY   Guarded platform around the cap of a tower/smock mill.

HARPS   Triangular segments of the grinding faces of millstones, containing lands and furrows.

HOLLOW POST MILL   A post where the drive is taken by a vertical shaft through the hollowed centre of the main post.

HOPPER   An open container to hold grain supported above the millstones.

HORSE   A wooden frame which supports the hopper over the stones.

KILN   A building, either separate from or attached to a mill for drying grain prior to grinding, especially in highland areas.

LANDS   Part of the grinding surface of millstones between the furrows.

LANTERN GEAR   An early form of gear with staves mounted between two discs or flanges.

LAUNDER   A trough carrying water on to a waterwheel.

LIFTERS   Cams driven into a shaft or stamps barrel to raise the stamps, which then fall by gravity.

LUCAM   A projecting gable enclosing an external sack hoist.

MEAL   Ground corn as it leaves the millstones.

NECK BEARING	The bearing supporting the neck or front of a windshaft.
OVER-DRIVEN	A mill where the millstones are driven from above.
OVER-SHOT WHEEL	A waterwheel where the water is carried over the top and the weight of the water turns the wheel.
PATENT SAILS	A form of easily regulated shuttered sail patented in 1807.
PEG MILL	Lancashire term for a post mill.
PELTON WHEEL	A form of impulse turbine developed in America in the second half of the nineteenth century.
PINION	The smaller of two wheels in gear.
PITCH-BACK WHEEL	A waterwheel where the water enters at the top but falls down the back of the wheel on the side of entry.
POLL END	The morticed end of a timber windshaft or an applied iron socket which holds the sail stocks.
POST MILL	A windmill of which the body containing the machinery and carrying the sails is turned into the wind about a vertical timber post.
QUARTERBARS	Diagonal struts supporting the post of a post mill.
RING GEAR	An iron toothed ring fixed to the shrouds or arms of a waterwheel from which the drive is taken.
ROLLER MILL	A nineteenth-century development in which iron or steel rollers were used to mill corn, superseding millstones.
ROUNDHOUSE	A circular or faceted building enclosing the sub-structure of a post mill.
SACK HOIST	A mechanism, usually power driven, to raise sacks into or through a mill.
SCOOP WHEEL	A driven wheel with raking floats to raise water in land drainage.
SHOE	An inclined wooden trough which is vibrated to allow grain to enter evenly into the eye of the stones.
SHROUDS	The outer rims of a waterwheel between which the buckets are fixed.
SHUTTERED SAILS	Spring or Patent sails, set by closing linked canvas or wooden shutters held in the sail frames.
SKIRT	Vertical weather-boarding around the bottom of the cap of a tower/smock mill or the base of a post mill body.
SMOCK MILL	A mill with a fixed body and moveable cap of which the tower is constructed of timber.
SPIDER	Coupling at the centre of a Patent sail assembly linking the shutter bars.
SPRING SAILS	An early form of regulated shutter sail introduced in 1772.
STAMPS	Vertical iron-shod timbers—later all iron—used as pestles to break up materials such as tin ore.
STAMPS BARREL	Horizontal cam-shaft with lifters to raise stamps.
STONE DRESSING	The skill of producing a cutting edge on the grinding faces of millstones.
SWEEPS	A term for sails, localised to Kent and Sussex.
TAIL	The rear of a post mill.
TAIL BOX	A compartment under the rear of the cap of some tower/smock mills which houses the winding gear for the cap.
TAIL POLE	A massive lever projecting from the rear of a post mill by which the mill was turned into the wind.
TAIL RACE	The water leaving a waterwheel.

TAIL WHEEL	The rear gear wheel on a windshaft, mounted behind the brake wheel.
THRIFT	Wooden handle into which, or on to which, bills are fixed for stone-dressing.
TIDE MILL	A watermill worked by water impounded in a pond when the tide is in and released to drive the wheel as it ebbs.
TOWER MILL	A windmill with a fixed body and moveable cap, of which the tower is of masonry.
TUN	The circular or faceted casing enclosing the millstones.
TURBINE	A form of water-driven prime mover developed in the nineteenth century which worked at much greater speeds than the conventional waterwheel.
UNDER-DRIVEN	A mill where the millstones are driven from below.
UNDER-SHOT WHEEL	A waterwheel where water enters at the lowest point and turns the wheel by its flow.
UPRIGHT SHAFT	The principal vertical drive shaft in a mill.
WALLOWER	The first driven gear in a mill, meshing with the brake wheel of a windmill or the pit wheel of a watermill.
WIND ENGINE	Windmills of which the sails are built in a circle and are usually mounted on skeleton-type towers.
WINDING GEAR	The gearing by which the cap of a tower/smock mill is turned into the wind.
WINDSHAFT	The principal drive shaft of a windmill which carries the sails at its outer end and is turned by them.